Electric Circuits: Misconceptions Clarified

Peter Sabato

October 2013

Peter Sabato   October 2013

ISBN-13:978-1492873921

ISBN-10:1492873926

REVISED VIEW

A simple circuit: A 10V voltage source in series with and a 10 ohm resistor and a lamp (not fixed resistance).

Typical approach when solving for the lamp voltage:

I have 10V, a 10 ohm resistor and a lamp in series so I must have 10V at the lamp to start off.

Clarification:

If the series circuit is open between the resistor and the lamp, there is a 10V potential difference at either side of the resistor and ground, with the current available to an additional load to be limited by the internal resistance of the 10V source at the 10V source/10 ohm resistor junction and the internal resistance of the 10V source plus the 10 ohm resistor at the open end of the 10 ohm resistor.

When the circuit is broken at the resistor/lamp junction, yes, there is 10V at the open end of the resistor and ground, but, when the circuit is closed, the lamp voltage is continuously changing (although in a narrow region measurable with a fluctuation) due to filament temperature, ambient temperature and even 10V voltage source stability and its dependence on internal temperature ambient temperature along with the 10 ohm resistors' resistance deviation due to internal temperature and ambient temperature. Additional variances include conductor dependency, gauge, composition and switch similar dependencies.

The voltage at the lamp is in flux with the total circuit. If the lamp is more efficient and produces a light (and/or infra-red) output and its internal resistance is higher, meaning less conduction, then yes, the lamp voltage increases to the 10V voltage source rather than by way of more conduction and more current draw, diminishing the voltage towards the ground or zero voltage level.

In electronics, often by the fact that a component is not going to draw abnormally high current, therefore almost allowing the applied voltage to become the component voltage, the maximum voltage rating of the component must be taken into consideration. This becomes more critical in capacitors involved in electric circuits that can easily be "over-voltaged" and thus destroyed.

The less the resistance in the lamp, the more the current through the lamp and thus the less the voltage across the lamp. In troubleshooting methods, someone may look for the voltage across the lamp and realize it is virtually zero, meaning a possible short at the lamp case between center and ground terminal, bypassing the current directly to ground, if in fact the 10V voltage source is still present.

In "pumping up" the current through the lamp, what would one expect at the 10 ohm resistor/lamp junction? The answer is that since more current is flowing through the lamp, and since the circuit is series, there is more current (same as in the lamp) through the 10 ohm resistor now. The higher current through the lamp "pumps up" the current through the 10 ohm resistor (up to the point of its circuit inherent natural limit of 10V/10 ohms, one ampere). This "pumped up" current action at the lamp increases the voltage at the resistor and is less "shared with the voltage across the lamp. The 10 ohm resistor now has most of the voltage across it. In stable designs, it is very important for one to take into consideration the maximum power that the resistor can handle, current X voltage of the resistor so as to not exceed the resistor's power (thermal handling capacity) in possible and probable circuit extremes which will cause the circuit to self-destruct. Since the resistor is less variant and more stable, measuring across it in any circuit condition can determine the status of the other components, specifically operating current, with the voltage across it divided by the resistance of the resistor itself, typically performed in current monitoring, current limiting and characterization of the lamp/load.

HYPED UP BY ADVERTIZED SPECIFICATIONS

An audio enthusiast reads that a certain amplifier produces 400 watts in comparison to another amplifier which produces 200 watts. A decision is made to purchase the 400 watt amplifier.

Clarification:

The enthusiast needs to understand that many methods of determining the output power can be made, with all of the methods correct, but, since there are many methods of calculating output power, he needs to adopt one method and transfer/ recalculate advertised specifications to conform to the identical comparison technique that revises all other power outputs to direct and proper comparison.

What if the manufacturer advertises his amplifier's power output in "peak to peak" power where another advertises in "peak" power while another advertises in RMS power? Well below any measured or heard distortion (total harmonic distortion THD) there needs to be one and only one method of comparison. While various ratings are true in the form that they were determined, and can be converted into another form, for example RMS to 'Peak to Peak" one, usually RMS power, is the calculation and reference of comparison, of choice. Measuring a high power audio output from an amplifier to a speaker, across a 0.01 ohm resistor with an isolated oscilloscope is a fast method to determine the power. Voltage over the 0.01 ohm resistor is the circuit current, seen and measured in RMS, peak and peak to peak formats, keeping one format, to compare real power, is the key.

Additionally look for select frequency maximum power vs. multi-frequency (real world music output) to understand the amplifier's capability. As more tones are amplified, the less the amplifier is able to keep the output from distorting, therefore, the power gain adjustment needs to decreased, to compensate, It is the multi-frequency, real world music, not single tone, that matters in an audio amplifiers amplifier's performance, except in most cases of the bass player using mostly single tones at any given time.

Someone looks at a vacuum cleaner and says "this one is 12 amps, it must be better than the one rated at 10 amps". Or this hair dryer is 1500 watts, it must be better than the one rated at 1200 watts.

Clarification:

Higher current does not mean better, in fact if compared in a laboratory, the researcher will become elated if the 10amp vacuum cleaner or the 1200 watt hair dryer produces the same usable thermal and mechanical power as the 12 amp vacuum cleaner or the 1500 watt hair dryer. Why? The goal of the designer is to use less power, (regardless of the measurement of device current or actual voltage X current, power rating in watts) resulting in higher efficiency, (output power in watts/consumed power in watts. This is the goal, higher efficiency to ultimately consume less power, less current, not behold the high current and power/wattage numbers in awe and be convinced that this device is better.

STATIC ELECTRICITY

Once, a manufacturing facility of InGaAs focal plane arrays I worked for in Princeton, was losing 90% of the extremely expensive devices in a start-up company that every dollar mattered in its survival. Every time I received the device for testing, one had to put on a clean-room jacket to make the transfer from clean-room to test laboratory. Every time in the removal of the jacket at the clean-room changing area, I felt a massive spark, static electricity that resulted from the creation of friction of the garment. This happened with others as well. I advised the manager that these events were happening and with a sharp imagination, placed humidifiers throughout the entire facility. This reduced destruction of the devices to a reasonable level. Today, there are static free garments, flooring and shoes available for static sensitive environments without the need for humidifiers.

Clarification:

Although massive expense is spent on a facility, the very return on investment ROI, is obliterated, if environmental conditions of the manufactured device are not taken into consideration and solved.

HIGH VOLTAGE WITH A GROUND STRAP

A technician working with high voltage told me of differing advice given to him when working with deadly voltages in a manufacturing environment, yet static free environment. One advice he received was to wear static conducting shoes even while working with high voltage. The other advice was opposite, do not work with static conductive shoes while working with high voltage. Which one was correct?

Clarification:

To save the device, work with static conducting shoes, but to save your life, do not wear static sensitive shoes as one should not be a path to ground through themselves.

Before static-free garments were mandatory, I always used to find a ground point on the lab bench and discharge and built-up potential differences, before continuing the experiment.

HERE IS YOUR LAB

complete with fluorescent lights, next to the room where they are re-polishing the floor with heavy equipment, on the wooden table with the temperature able to change from 60 degrees F to 80 degrees F. A vacuum cleaner can be used if you need it. It should be great for precision work.

Clarification:

No electromagnetic fields of 60HZ propagating from lighting transformers, no transients on the laboratory measuring equipment power line, no static generation or "capacitive holder of a potential" such as wood, no temperature difference at all. Most experiments always are void of ambient temperature conditions that become difficult to re-find that original experiment test temperature in solving contradicting data when duplicating experiments. No breezes, even paper movement can produce friction against paper in the lab, and cause wide variations in measured results. Vacuum cleaner, have you ever felt the static discharge from a vacuum cleaner after turned off, made by friction of air? (It is lightning generated by the fierce air currents, extremely dangerous to electronic circuits).

THESE ARE THE RESULTS, THE LAW

Clarification:

The scientist always looks to find more and more variables in an experiment, so as to see their effect, limit them in studied for experiment replication and to further understand the science that builds up what we think we know so far.

A motor is rotating at 300RPMs, 28 Volts is applied, there are two amperes flowing, however the next day, when the experiment is repeated, there is only one amp flowing for the same result. What was the difference? The difference can be found with as many parameters of the test identified and documented. It does not matter how insignificant any parameter may seem, electronics is a science whose laws are bound to unchanging principles, with deviations due to oversights in our record keeping.

What was the load on the output shaft of the motor? What was the humidity in the room?, the temperature? Did the measuring equipment basic range rise and fall with transients from next room heavy equipment injection on the line, susceptibility and sensitivity to these transients? Did the transients inject the EMF onto the wiring of the measuring equipment, which the measuring equipment measured, but was generation of voltage transients to non-shielded wiring open long wires, ripe for injection of EMF.

Once, I noticed an oscilloscope on, on the bench, not connected to any circuit, and still reading a signal. What could that signal be? I calculated 1/time to find the frequency of the signal, it was about 1.2 MHZ.

I could see the signal moving, as if dynamic, not repetitive but random. It looked like an AM signal at that oscilloscope setting when I turned up the v/cm range. This is AM radio as I turned on a small radio in the next room and found perfect matching with the signal. Very powerful EMF! Interesting and absolutely essential in the relevance of what signal is being isolated for examination in an experiment, hence RF shielding rooms exist.

I have seen the effect of ambient fluorescent or incandescent light on sensitive electronic circuits, even without a photo diode, photo-resistor or photo-cell involved. Why would the voltage on a seemingly simple experiment change when the lights are on? There is absolutely, photo-electric generation on various substances, causing measurable changes, exposed metallic surfaces connected in series, or connected directly to a voltmeter shows the effect of light. Thermal effects from the light in contact with various components change the experiment with a great degree.

Meters fluctuating in LCD read-out results, unstable at times, have been questioned by assistants in laboratories. Whether the EMF of the circuit under experimentation interferes with the processor of the meter at times, from adjacent proximity, or the processor of the meter injecting itself onto susceptible, mostly open, long, untwisted wires of the experiment, either way has happened before. What about the long, untwisted leads of the meter? Absolutely, this is an antenna and tuned circuit for unwanted signal altering the real experiment results.

LET'S PUT A ZENER DIODE ON IT

to squelch the increase in voltage and isolate the voltage sensitive section.

Clarification:

Some circuits can be saved from voltage increase and although a zener diode mode of use "presumes" that things are OK, many circuits, such as laser diodes and LED displays are sensitive to extremely small variations in applied voltage (causing very big changes in conducted current) enough to promptly cancel any zener diode savior program. In laser diodes, an entirely different format of operation is essential for operator regulation of the output emission power. No longer can voltage change provide laser diodes smooth regulation but, current control, with a constant voltage applied be the superior method.

In LCD illumination regulation, I was once given the opportunity to design a satellite DC power supply test equipment box (after my request was granted). I noticed that the system had to check the presence of a varying voltage input, (14V to 38V) without changing the intensity of the LED used in its identification of presence, regardless of voltage. Without designing a trigger from the presence of the input voltage to an on/off state regulated intensity indicator, which any loss of this more complex circuit had to be taken into account because efficiency measurements were critical, and complexity of the box wasn't extremely necessary, I put a zener diode in parallel with the LED(s) (in series with a 10K resistor) so that despite the voltage range change, the LED(s) would not under or over glow and possibly not indicate presence or self-destruct, by overvoltage. I found that these zener diodes did not help very much at all. I changed the design to use current limit diodes CLDs. After an in-series placement of this device with the LED(s), absolute unchanged intensity occurred regardless of applied voltage, a first design of test equipment using CLDs at the satellite design facility in Princeton. CLDs are really magic.

In a related and critical laser system design at a company in Edison, I noticed that $20,000 laser modules were being blown up every time they were installed into a system, there would be blame on the production side vs. the system side as to who was at fault. I decided to isolate the power supply to the system laser module even though it was an off-the-wall, out-side-the-box idea. What I found with an oscilloscope at turn-on was unbelievable. Upon laser enable (power supply turn-on) (not even foot switch laser-on delivery), there was a spike of 100V over the rating of the laser module. Soft-start needed to be looked at and when redesigned, the power supply ramped up to the same operating voltage as before, but without the 100V transient. All laser diode modules from then on were fine.

HIGH VOLTAGE ISOLATION MATERIAL

It's OK, we can put a layer of that to off-stand 1000V.

Clarification:

In theory, one layer of insulation to off-stand high voltage sounds great until there is a metal shard within the material which discharges the entire system into a 911 emergency. Account for the addition of the unexpected. I saw a YAG/CO2 laser blown to bits because of a tiny metal shard discharging the entire system. Always place an extra layer of material between. Better yet, use glass or plastic isolated case FETs/IGBTs. Save the system, especially if it is a human body contact medical device.

IT WILL BE JUST FINE NOW

Clarification:

Temperature test everything. Low, high at maximum output, maximum voltage input. Don't let someone else do it for you after a failed system.

DE-IONIZED WATER ISOLATION NEVER A PROBLEM

Clarification:

If at all possible, get away from DI water, DI water wants to be ionized, it strives quickly to reach this state, removing insulation from high voltages, endangering any operators and maintenance-intensive.

In integrated circuit packaging, DI water is only a theory, when involved with test, increases in levels of conductivity inject precious micro amps of leakage to case pin-outs to each other Vdd and ground.

FETs ARE ALWAYS SURPERIOR TO IGBTs

Clarification:

FETs are amazing devices, beautiful to work with and see the operation in the linear region. A quick check of N-FETs include small, with a handheld meter, on-the spot checks such as biasing the gate positive to source just to see if the drain to source is being conducted, checking the body drain diode, if it is still there, turning off the FET by reversing the positive bias on the gate to source to negative. A FET can be used accurately for adjustable current loads, provided they are not very high powers. However, a FET is not the best device for high power dynamic loads. The IGBT locks in the Vcollector voltage to 1.5V maximum, regardless of current through it. A clamping switch, not allowing the Vd X Id (Pd) (thermal dissipation) to climb to 5V, 10V X the current through it, throwing the FET into its linear region, even though it is supposed to be clamped, burning up quickly. The power dissipated through a "clamped" FET is insanely high, do not use FETs for high power, after consideration to parallel them has been tried. An IGBT promptly locks down the current into a 1.5Vmaximum limiting losses and providing predictable results. I designed a load for a 3 phase test (manufacturing) and used FETs first, it's IGBTs all the way after that mistake.

WHY DOES THIS ONE DRAW SO MUCH POWER?

Clarification:

A circuit may be designed and work, but the next serial number of the same devices draw hugh current and doesn't work. Nine times out of ten, it is oscillation. Check the output with an oscilloscope, is there loading to ground when there shouldn't be? Couldn't hear it because the oscillation was in the ultrasonic range (audio amplifier, for example) An entire book can be written about this, but stability (phase/gain margin) is important. By the way, what is your system doing to the power line, adding transients? Very important. Under scrutiny, these parameters are being checked more and more.

IT'S CONNECTED, MUST BE SOMETHING ELSE

Clarification:

Connections in electronics are most prone to the degradation of any system. I have seen this time and time again. As more current is needed, connections become more important.

Clean, tight connections, ones which are not so bad that they heat up their surroundings, melt their plastic housings and cause fires. I saw a system that consumed 150 amps from the output of a transformer. In 200 out of 250 systems, the transformer output was lightly bolted to a pcb. In these systems, all 200 had visibly burned connections at the transformer pcb junction, some even with hardware that evaporated from the mating junction, all reduced the current needed to supply the rest of the system down-line, to hazard mode, these rectified rails of Vdd to the high power 10KW RF amplifier provoked diminished performance, and exponentially degraded the connections with every minute of use. The return rate from the customer was astronomical. I recommended a torque setting unheard of for the manufacturing process, there were no more burned connections and no more returns due to the carbonization and oxidation, lack of conduction and massive overheating of the transformer pcb junction anymore.

THE CIRCUIT IS ON WHEN IT SHOULD BE OFF

Clarification:

A long time ago, discovered in tube and transistor circuits, I discovered that a hand touching the grid or base or even gate of a device will absolutely turn it on. Not only turn it on, but induce the 60HZ EMF from the adjacent environment into the system. This is most noticed when the center pin of an audio amplifier is touched, the speakers max out with an insane harsh volume of 60HZ. But mostly, during the probing of sensitive circuits with meters, please don't have your hand on the lead, it will trigger an on state (non-ground node)

It is bad enough that many times the meter itself influences the circuit and may turn on the system with the antenna-like leads that just desire to inject noise into the system.

www.ingramcontent.com/pod-product-compliance
Lightning Source LLC
Chambersburg PA
CBHW071604170526
45166CB00004B/1797